Air, Water and Soil Pollution Science and Technology

Perchlorate Formation in Electrochemical Water Disinfection

AIR, WATER AND SOIL POLLUTION SCIENCE AND TECHNOLOGY

Additional books in this series can be found on Nova's website under the Series tab.

Additional E-books in this series can be found on Nova's website under the E-books tab.

WATER RESOURCE PLANNING, DEVELOPMENT AND MANAGEMENT

Additional books in this series can be found on Nova's website under the Series tab.

Additional E-books in this series can be found on Nova's website under the E-books tab.

AIR, WATER AND SOIL POLLUTION SCIENCE AND TECHNOLOGY

PERCHLORATE FORMATION IN ELECTROCHEMICAL WATER DISINFECTION

M. E. HENRY BERGMANN
TATIANA IOURTCHOUK
WIDO SCHMIDT
GABRIELE NÜSKE
AND
MICHAELA FISCHER

Nova Science Publishers, Inc.
New York

Copyright © 2012 by Nova Science Publishers, Inc.

All rights reserved. No part of this book may be reproduced, stored in a retrieval system or transmitted in any form or by any means: electronic, electrostatic, magnetic, tape, mechanical photocopying, recording or otherwise without the written permission of the Publisher.

For permission to use material from this book please contact us:
Telephone 631-231-7269; Fax 631-231-8175
Web Site: http://www.novapublishers.com

NOTICE TO THE READER

The Publisher has taken reasonable care in the preparation of this book, but makes no expressed or implied warranty of any kind and assumes no responsibility for any errors or omissions. No liability is assumed for incidental or consequential damages in connection with or arising out of information contained in this book. The Publisher shall not be liable for any special, consequential, or exemplary damages resulting, in whole or in part, from the readers' use of, or reliance upon, this material. Any parts of this book based on government reports are so indicated and copyright is claimed for those parts to the extent applicable to compilations of such works.

Independent verification should be sought for any data, advice or recommendations contained in this book. In addition, no responsibility is assumed by the publisher for any injury and/or damage to persons or property arising from any methods, products, instructions, ideas or otherwise contained in this publication.

This publication is designed to provide accurate and authoritative information with regard to the subject matter covered herein. It is sold with the clear understanding that the Publisher is not engaged in rendering legal or any other professional services. If legal or any other expert assistance is required, the services of a competent person should be sought. FROM A DECLARATION OF PARTICIPANTS JOINTLY ADOPTED BY A COMMITTEE OF THE AMERICAN BAR ASSOCIATION AND A COMMITTEE OF PUBLISHERS.

Additional color graphics may be available in the e-book version of this book.

Library of Congress Cataloging-in-Publication Data

Perchlorate formation in electrochemical water disinfection / authors, M.E. Henry Bergmann ... [et al.].
  p. cm.
 Includes bibliographical references and index.
 ISBN 978-1-61209-690-2 (softcover)
 1. Water--Purification--Perchlorate removal. 2. Perchlorates--Electric properties. 3. Electrochemistry, Industrial. I. Bergmann, M. E. Henry.
 TD427.P33P46 2011
 628.1'662--dc22   2011008475

Published by Nova Science Publishers, Inc. † New York

Contents

Preface		viii
Abstract		ix
Symbols and Abbreviations		xi
Acknowledgement		xiii
Chapter 1	Introduction	1
Chapter 2	Occurrence of Perchlorate in Electrolysis	5
Chapter 3	Experimental Details	7
Chapter 4	Results and Discussion	13
Chapter 5	Discussion of Probable Mechanisms	27
Chapter 6	Conclusions	39
References		41
Index		47

PREFACE

Electrochemical water treatment is a simple method to generate disinfecting agents. Several companies offer special cell technologies for water disinfection. It has been found that besides active chlorine for disinfection, by-products such as chlorate and perchlorate may be formed. Systematic studies using laboratory cells, semi-technical and technical cells confirm this. Mixed oxide and boron doped diamond anodes were used in laboratory-scale and semi-technical experiments which were conducted under drinking water conditions at temperatures between 10 and 30°C and at current densities between 50 and 500 A m^{-2}. The results of these studies show a perchlorate formation potential. This new book presents and discusses current research in the study of perchlorate formation in electrochemical water disinfection.

ABSTRACT

Electrochemical water treatment is a simple method to generate disinfecting agents. Several companies offer special cell technologies for water disinfection. It has been found that besides active chlorine for disinfection, by-products such as chlorate and perchlorate may be formed. Systematic studies using laboratory cells, semi-technical and technical cells confirm this. Mixed oxide and boron doped diamond anodes were used in laboratory-scale and semi-technical experiments which were conducted under drinking water conditions (chlorine concentration up to 250 mg L^{-1}) at temperatures between 10 and 30°C and at current densities between 50 and 500 A m^{-2}. The results of these studies show a perchlorate formation potential. Measured perchlorate concentrations are often in the mg L-1 concentration range, i.e. thousandfold higher than recommended by US water authorities. Several technical cells, which were in use and equipped with mixed oxide electrodes were tested with respect to perchlorate formation in the range of 1.8 - 56.7 g [Cl$^-$] L^{-1}, 0.1 - 28.4 g L^{-1} of active chlorine and at currents between 6 A and 150 A. The highest perchlorate concentration, which was found in these examinations, was 9 mg L^{-1}. Risk factors for enlarged occurrence of perchlorate are higher current densities (electrode potentials), longer treatment time (high specific charge flow), water matrixes with relative low ionic strength and the use of electrodes having a high oxidation potential. Extended studies show that the problem is not only related with drinking water treatment. All aqueous systems containing chloride, hypochlorite or chlorate ions are able to form perchlorate during electrolysis under certain conditions. Therefore, the certification of the cells is recommended when used for drinking water disinfection to avoid health risks for consumers.

SYMBOLS AND ABBREVIATIONS

AC	Active chlorine, mg L^{-1}
ads	Adsorbed state
BDD	Boron Doped Diamond
C_{eff}	Effective chlorine generation, mg[Cl] per mg[Cl$^-$]
c_{sp}	Specific perchlorate formation, µg[PC] per mg[AC]
DPD	N,N-diethyl-p-phenylendiamine
DSA	Dimensionally Stable Anode
EPA	Environmental Protection Agency
F	FARADAY constant, 96486 As mol^{-1}
FAC	Free Available Chlorine, mg L^{-1}
I	Current, A
K	Rate constant, M^{-1} s^{-1} for second order reactions
MO	Mixed Oxide
n	Number of moles
PC	Perchlorate
RNO	*N,N*-dimethyl-4-nitrosoaniline
rds	Rate determining step
rpm	Revolutions per minute
SHE	Standard Hydrogen Electrode
SMSE	Saturated Mercury Sulphate Electrode
t	Time, s
Q_{sp}	Specific charge flow, Ah L^{-1}
°dH	German degree of water hardness

Kinetic (overall) reaction constants were taken from literature sources including the Radiation Chemical Data Center.

ACKNOWLEDGEMENT

The work was supported by BMBF/AIF FKZ 1721X04 and DVGW (project no. W4/05/06). The authors wish to thank Johanna Rollin, Christine Hummel, Karsten Kresse, Christian Czichos, Andreas Rittel (Anhalt University), Karel Bouzek and Roman Kodym (VSCHT Prague), and Savas Koparal (Anadolu University Eskisehir) for technical, analytical and mathematical support.

Chapter 1

INTRODUCTION

There is no clear definition of the term *Electrochemical Water Disinfection*. Also, the term 'electrochemical treatment' is sometimes used for processes, which are not based on electrochemistry. Several chemical disinfectants can be electrochemically generated and stored (chlorine gas and chlorinated lime solutions). But these technologies are not within the scope of this paper. However, some results of the examination of different $HOCl/OCl^-$ stock solutions, which were obtained after the electrolysis of a concentrated brine solution, are added to this chapter. In all other cases, *Electrochemical Disinfection* is used in the meaning of disinfectant generation when more or less natural water is passed through an electrochemical cell. Users understand the process as the production of Active Chlorine (AC as sum of dissolved chlorine + hypochlorous acid + hypochlorite) or Free Available Chlorine (FAC) to kill microorganisms. In the past it was claimed that the killing of microorganisms during electrolysis is a result of their contact with the electrode surface, the strength of the electrical field or not further defined implosions. A first critical discussion concerning the mechanisms, which lead to cell death, was given in [Bergmann et al, 2001]. Since the middle of the 20th century Reis and co-workers have published several papers about the method calling it Anodic Oxidation [Reis, 1951, Reis and Henninger, 1953, Reis, 1976, Reis, 1981, Gutknecht et al., 1981].

The discussion of electrochemical disinfection is accompanied by serious problems in understanding because of the large variety of materials, cell constructions and treatment conditions. There are numerous brand names, for example *Anodic Oxidation, Electrochemical Water Sterilization* or *Activation, Low Amperage Electrolysis* and *In line Electrolysis*. Additionally, some cell producers refer to it as *Tube Electrolysis* because they prefer a tube construction design for the electrolyser.

Chlorine generators of different types and size are applied in waterworks, social facilities e.g. in hospitals, senior hostels and sports facilities. Figure 1 demonstrates schematically three typical cells. Figure 1a shows a simple flow-through cell. The generated disinfectants continue to kill pathogens even many minutes after having passed the cell. Mostly, there are many pairs of electrodes combined in a so-called electrical bipolar or monopolar connection for higher disinfectant production. The majority of electrode materials is able to generate AC, for example graphite, activated titanium and platinum. Doped diamond is a relatively new electrode material [Fujishima et al., 2005]. In contrast to variant (a) in Figure 1, variant (b) shows a divided cell with additional catholyte handling. The use of a separator reduces the amount of gases in the water and possible oxidant losses due to re-reduction at the cathode and avoids formation of H_2/O_2 mixtures. But it also increases the cell voltage. Some companies use the term *Diaphragmalysis* for this variant that is widespread for the so-called onsite production of disinfectants which means that a brine solution is passed through the electrolyser. The product of the electrolysis is a $HOCl/OCl^-$ stock solution, which is added to disinfect water. Figure 1c shows a discontinuous treatment regime, typical for warm water systems, submarine water supply etc. Producers recommend the addition of sodium chloride salt if the water contains only a few mg L^{-1} of chloride. (mg L^{-1} or ppm are used as concentration units in practice and regulations as well as in this paper.) The efficiency of AC production significantly depends on the chloride concentration. Often, no active chlorine can be found if the chloride concentration is lower than 10-20 mg L^{-1}.

There is no doubt in the disinfection ability of electrochemical water treatment methods. State of research was summarized in [Martinez-Huitle, 2008, Bergmann 2009a].

In previous publications the formation and occurrence of by-products in disinfection electrolysis were not considered. However, new studies show a clear potential of the formation of inorganic by-products. Perchlorate is a representative of these.

The intention of this chapter is to give an overview of the peculiarities of perchlorate formation at different electrode materials using lab-scale, semi-technical and technical cells. As can be seen later the formation potential of perchlorate is high under conditions of favored radical generation at anodes.

Chapter 2

OCCURRENCE OF PERCHLORATE IN ELECTROLYSIS

Perchlorate, which is a substance of high oxidation state, is synthesized from species of lower oxidation state by energy input. Electrochemical methods are one way to accomplish this. The first synthesis was reported in the beginning of the 19th century. Platinum, lead dioxide and mixed oxides were typical anode materials for perchlorate production. Literature on electrochemically generated perchlorate has been published since many decades [Kirk-Othmer, 1979, Kuhn, 1971] but the reaction mechanisms are still not known in detail. Chlorate seems to be an important intermediate in perchlorate electrolysis [Udupa et al., 1971]. Djordjevic and co-workers [Djordevic et al., 1973] suppose OCl^*, ClO_2, ClO_2^- as intermediates for ClO_3^- formation electrolyzing hypochlorite solution. Besides chlorate also chlorite and bromate can occur during the electrolysis. The drinking water guidelines of the WHO recommend maximum concentrations of 0.7 mg L^{-1} for chlorite and chlorate and 0.01 mg L^{-1} for bromate.

The first occurrence of perchlorate in an electrochemical drinking water treatment procedure was reported by Jackson, who found perchlorate obviously as a result of electrochemical corrosion protection in water storage tanks [Jackson, 2004]. The effect was systematically studied by Tock et al. [Tock et al. 2004] using silicon iron and coated titanium as anode material. A new paper of Jung and co-workers presents chlorate and perchlorate formation on a platinum anode [Jung et al., 2010]. Recently, it was found that boron doped diamond electrodes

(BDD) may produce perchlorate if chloride, hypochlorite or chlorate solutions, synthetic or real drinking water are electrolysed [Bergmann et al., 2007a, Bergmann et. al, 2007b, Bergmann et al., 2009b].

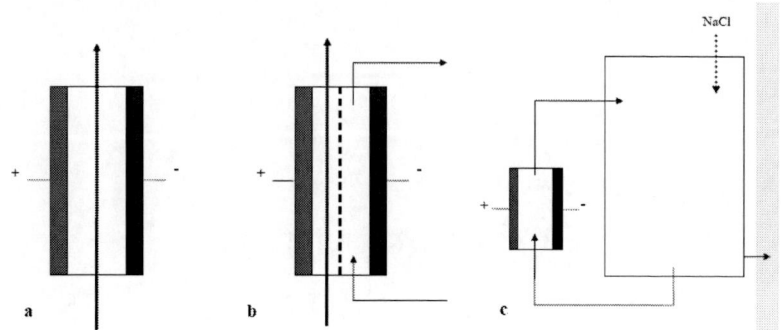

Figure 1. Selected technological cell designs for electrochemical water disinfection (a- undivided cell with flow-through regime for main or by-pass water streams, b- divided cell with additional catholyte, c- discontinuous treatment regime with optional NaCl addition.

Results were confirmed by Polcaro and co-workers [Polcaro et al., 2008]. Stanford and co-workers found perchlorate in mg L^{-1} concentration range in hypochlorite stock solutions [Stanford et al., 2009]. In general, hypochlorite solutions are prepared by discontinuous chloride electrolysis. The occurrence of chlorate is a well-known fact but also the occurrence of perchlorate has already been reported in the middle of the 20th century [D'Ans and Freud, 1957]. D'Ans and co-workers found out that perchlorate is a by-product in the reaction of hypochlorite/hypochlorous acid with chlorate. They explained the formation by a fast formation of an activated complex [HCl_2O_4]- that may slowly react to perchlorate, H^+ and Cl^-.

Today, a safety value for perchlorate is mainly discussed in the USA [Gu and Coates, 2006, Sellers et al., 2007]. The US EPA discussed an interim health advisory level of 0.015 mg L^{-1} for perchlorate in drinking water [EPA, 2008], which may serve as an advisory rule for other countries, as well.

Chapter 3

EXPERIMENTAL DETAILS

LABORATORY-SCALE AND SEMI-TECHNICAL CELLS FOR WATER ELECTROLYSIS

Electrolysis experiments were mostly run on a cell with rotating BDD anode as described in [Bergmann, 2005,Bergmann, 2009a, Bergmann et al., 2009b] and illustrated in Figure 2a. The circular plate anode had a diameter of 35 mm and was located 4-5 mm above an expanded mesh cathode of the same diameter (*Magneto Chemie*, molaric ratio Ir:Ru = 50%/50%, ratio of active surface to projected surface of 1:1). During the experiments two anode materials were used. At first a boron doped electrode (*Condias*, 2000-4000 ppm) with a thickness of about 1-2 μm [Bergmann et al., 2009a] and secondly an electrode made of IrO_2/RuO_2 mixed oxide (MO, producer *Magneto Chemie*) where the molaric ratio of Ir:Ru=1 was used. Anode shaft and cathode contacts were isolated by *Viton*. The volume of the electrolyte was usually 100 mL. The cell was an open dark glass beaker thermostated by a *Lauda RM6* cryostat. The current was adjusted using a *Statron 2252.2* rectifier.

Two half-cells with embedded BDD electrodes (*Condias*) were used in discontinuous experiments shown in Figure 2b. The electrode dimensions in mm were: 2x20x50. Both electrodes, which were produced for water treatment, had a boron doped layer as described above. The interelectrode distance was 4 mm. The water was recirculated (*Verder* centrifugal pump) out of a stirred dark 1.6 L glass beaker through the cell and a heat exchanger inside a cryostat bath to keep a constant temperature of 20°C. A *CSEM* cell with 2 bipolar electrode pairs (BDD on silicon, diameter 90 mm) shown in Figure 2c was tested with respect to perchlorate formation.

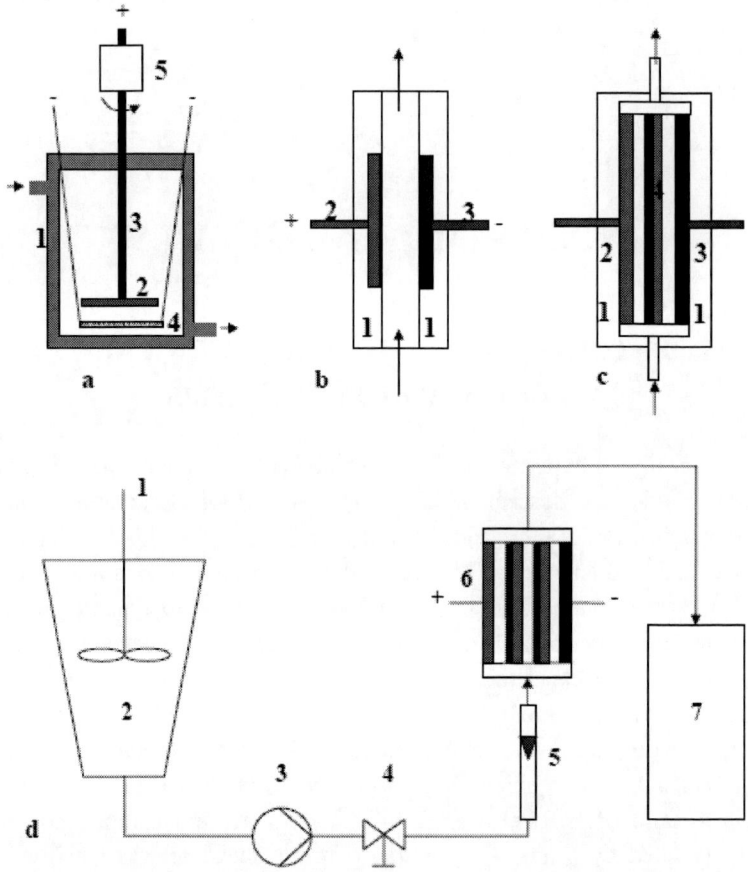

Figure 2. Cells used in the experiments: a- discontinuous cell with rotating disk anode, 1 thermostated beaker, 2 anode, 3 isolated shaft, 4 extended mesh cathode, 5 rotating contact; b- discontinuous cell with parallel plate electrodes, 1 half-cells, 2 embedded anode, 3 embedded cathode; c- bipolar cell with diamond electrodes, 1 cell housing, 2 contacted anode, 3 contacted cathode, 4 bipolar electrode; d- stand with bipolar MO stack cell, 1 stirrer, 2 300L container, 3 centrifugal pump, 4 valve, 5 rotameter, 6 cell, 7 container.

For the experiments with continuous operation mode and varying chloride concentration 300 L of drinking water with adjusted chloride concentration were stored in a plastic container and pumped continuously by a centrifugal *Grundfos JP5-B -B -CVBP* pump through a rotameter and the electrolyser. The inlet water temperature was between 16 and 18°C, outlet temperature was nearly 20°C.

In experiments shown in Figure 2d, the BDD cell was replaced by a bipolar stack cell consisting of 7 MO electrode pairs in a water filter housing (*Dr. Rittel GmbH*). The electrode area was 149 cm² for each electrode. The IrO_2/RuO_2 electrodes were supplied by *DeNora Germany* with a molaric Ru to Ir ratio of nearly 3.

All experiments were performed in constant current mode.

Polarization curves were studied using an *EG&G* potentiostat model 283 together with a rotating disc electrode stand (model 616) in IR compensation mode. The anode was a 1cm² BDD disc (*Condias*) on niobium. A mercury oxide reference electrode (*Sensortechnik Meinsberg*) in an outer beaker connected with a salt bridge containing 0.25 M NaOH was used to measure the potential. The cathode was a 15 mm x 30 mm platinum sheet.

Conductivity and pH were measured using *WTW* instruments (*Cond 340i, pH 340)*. Samples were taken at different time intervals and were immediately analyzed. Every experiment was performed at least twice. Hypochlorite formation was measured using the UV spectrophotometers *Specol 1200* or *Specord 40* (*Analytik Jena*) and DPD test 017 (*Macherey&Nagel, Nanocolor 100D*). Ion analytic was preferably performed by HPLC (see below).

Chemicals

Water for synthetic electrolytes was produced using a *Seralpur Pro 90 CN* deionization unit (conductivity<0.1 µS cm⁻¹). Chemicals had a purity of at least higher than 99.9%. The substances Na_2CO_3 (*Roth*), NaCl, NaOH, $Ca(OCl)_2$ (*Fisher Scientific*) and $NaClO_2$, $NaClO_3$ (*Lancaster*) had a purity of 99.99%, and NaCl and Na_2SO_4 (*Chempur)* had apurity of 99.999%. Some of the used chemicals (Na_2ClO_2, $NaClO_3$) were partially treated two to three times by fractional crystallization at different temperatures. Also regional drinking water was used which can be characterized by the following main parameters: chloride 42 - 46 mg L⁻¹, sulphate 145 - 165 mg L⁻¹, nitrate 10 - 13.9 mg L⁻¹, pH = 7.4 - 7.9 and water hardness 19.6 °dH.

Ion Analysis

For the comparability active chlorine concentration was checked in some experiments by UV spectroscopy after adding 25 µL 2M NaOH to the sample for adjusting it to pH 11.5. Chlorate and perchlorate were analyzed using a *Metrohm Metrosep Dual 4* column [Bergmann et al, 2007b] allowing perchlorate detection below 1 µg L^{-1}. Because the peaks of chloride and hypochlorite overlap, ion chromatography analysis was carried out twice - with and without chemical hypochlorite elimination. The corresponding Cl^- and OCl^- concentration values were calculated solving a system of mathematical equations. Nitrite, nitrate, chloride, hypochlorite, chlorite, chlorate and sulphate were analyzed by ion chromatography (*Knauer/Alltech* system with *Novasep A-2* anion column and electrochemical detector). Experiments were repeated at least once.

TECHNICAL CELLS FOR BRINE ELECTROLYSIS

Five types of commercially available chlorine generators were tested using a specified monitoring program in an examination done by the TZW. Several technical details and information concerning the operating conditions of the different cells are summarized in Table 1.

In general, three basic types of chlorine generators are available: Divided or undivided cells producing hypochlorite solution and divided cells generating chlorine gas, which is dissolved in sodium hydroxide afterwards. The operating conditions of the different generators covered a wide range, e.g. the cell current was varying from 6 up to 150 A and the pH-value of the generated product was measured from 2 up to 10.

The brine of three generators was adjusted to 280 g L^{-1} chloride and two cells were operating with 330 g L^{-1}. However, the actual chloride concentration in the cells before the electrolysis was much lower and covered a wide range (from 1.8 up to 56.7 g L^{-1}). Therefore, the concentration of produced active chlorine was determined in the range of 0.1 up to 28.4 g L^{-1}. Information concerning the electrode material which belongs to internal knowledge of the companies was very poor.

Table 1. Technical details of the tested generators

	Type	Brine [g L^{-1} Cl$^-$]	Cl$^-$ in cell [g L^{-1}]	Cell current [A]	Q_{sp} [Ah L^{-1}]	pH-value	Electrodes*
A 1	divided cell generation of hypochlorite solution	280	54.4	150		9-10	anode: titanium coated with ruthenium dioxide cathode: titanium coated with nickel alloy
A 2		280	54.4	150	6.47	9-10	
A 3		280	54.4	150		9-10	
B 1	divided cell generation of hypo chlorite solution	280	17	16		2.8-3.4	anode and cathode: mixed oxide
B 2		280	17	16	0.29	2.8-3.4	
B 3		280	17	8		2.8-3.4	
C 1	divided cell generation of hypochlorite solution	330	1.8	15	0.4	2-3	anode and cathode: mixed oxide
C 2		330	1.8	15	0.4	2-3	
C 3		330	1.8	30		2-3	
D 1	divided cell generation of chlorine gas	330	200	50		7-8	anode and cathode: mixed oxide
D 2		330	200	50		7-8	
E 1	undivided cell generation of hypochlorite solution	280	56.7	6	5.0	8-9	anode and cathode: titanium coated with mixed oxide
E 2		280	56.7	6		8-9	
E 3		280	56.7	6		8-9	

* Internal company knowledge.

PERCHLORATE ANALYSIS IN EXPERIMENTS USING TECHNICAL CELLS

The ionic by-products of the type XO_n^- were analyzed by ion chromatography using DIONEX equipment. The analysis method for perchlorate was optimized with regard to the determination of traces in concentrated brine solution.

The principle of this method is based upon the pre-separation of the brine matrix by column switching and the pre-concentration of the perchlorate ion on a special *IonPac Crypton* column (Figure 3). In addition, the concentration of chlorite, chlorate and bromate was analyzed using the procedure according to DIN EN ISO 15061 [DIN, 2001].

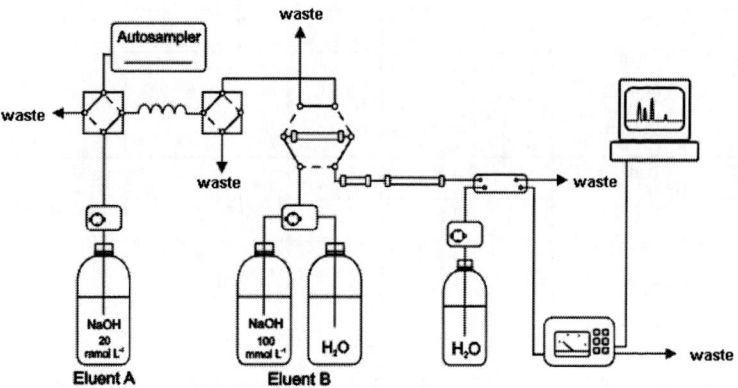

Figure 3. Scheme of the ion chromatograph including column switching and preconcentration of perchlorate in brine.

Chapter 4

RESULTS AND DISCUSSION

EXPERIMENTS IN LABORATORY-SCALE AND SEMI-INDUSTRIAL CELLS

Polarization

The occurrence of species generated in an electrochemical electrode process depends on the electrode potential applied. In the discussion of this chapter only anodic processes are considered because chloride has to be oxidized to produce higher-valent species. In general, also cathodic reactions may produce oxidants for instance H_2O_2 from dissolved oxygen or hydroxyl radicals from hydrogen peroxide but their contribution to the oxidation of water constituents is marginal:

$$O_2 + 2H^+ + 2e^- \rightarrow H_2O_2 \tag{1}$$

$$H_2O_2 + e^- \rightarrow {^*OH} + OH^- \tag{2}$$

A species conversion at an electrode surface is based on electron transfer and/or a reaction between species, at least partially generated by electron transfer reactions. The formation of perchlorate must include steps of water decomposition because perchlorate includes oxygen atoms. Oxygen is yielded by anodic water decomposition, which is mostly described by a reaction scheme given as follows:

$$OH^- \rightarrow OH^-_{ads} \qquad (3)$$

$$OH^-_{ads} \rightarrow {}^*OH + e^- \qquad (4)$$

or

$$H_2O \rightarrow H_2O_{ads} \qquad (5)$$

$$H_2O_{ads} \rightarrow OH_{ads} + H^+ + e^- \qquad (6)$$

$$2\,OH_{ads} \rightarrow O_{ads} + H_2O \qquad (7)$$

or

$$OH_{ads} \rightarrow O_{ads} + H^+ + e^- \qquad (8)$$

$$2O_{ads} \rightarrow O_2 \qquad (9)$$

This mechanism is applied to mixed oxide electrodes, platinum and PbO_2 anodes. In case of BDD anode the adsorption of hydroxyl radicals is considered as very weak (physically; please see discussion in section 5.1).

It is accepted that radicals such as the hydroxyl and the oxygen radical are temporarily formed. In addition to reactivity effects, it is a question of probability to what extent these radicals react with other compounds if those are available at the electrode surface. Chlorine components could be formed by this way.

Furthermore, one can expect that in case of high active site density on the electrode (and at low concentration of possible reaction partners) reactions as mentioned above (Eqs. 3-9) predominate, whereas under opposite conditions additional reactions are more probable. Thus, on BDD anodes with relatively low density of active sites, the perchlorate formation rate could be much higher than on mixed oxide and other materials. This prediction is confirmed (see results in section 4.1.2). So it becomes more apparent why in general at mixed oxide anodes perchlorate can be found although the anode potential is lower than the oxidation potential of, for example, hydroxyl radicals (Figure 4). At higher current density the curve is approaching an anodic potential of 1.5

V (SHE). Polarization curves were also presented in [Bergmann and Koparal, 2005]. A detailed discussion of polarization curves is not within the scope of this chapter.

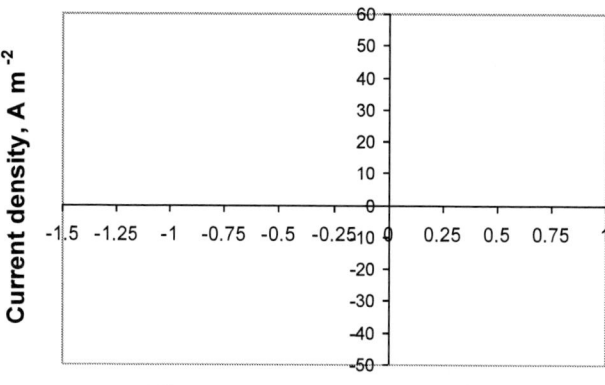

Figure 4. Anodic polarization curve on rotating mixed oxide electrode in drinking water (IrO_2/RuO_2 anode, rotation rate 1500 rpm, 247 mg [Cl^-] L^{-1} in tap water, 20°C; scan rate 1 mV s^{-1}, reference electrode: Saturated Mercury Sulphate Electrode at 0.640 V (SHE), IR compensation).

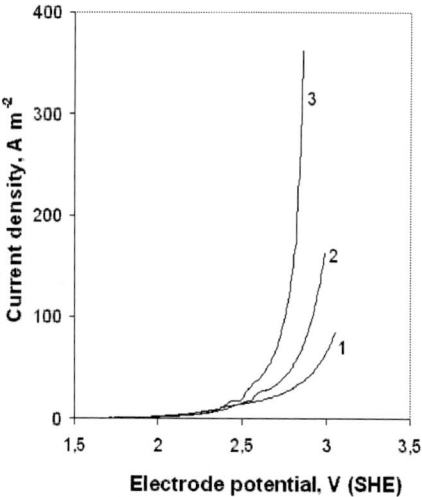

Figure 5. Polarization curves on rotating BDD anode in NaCl solution (chloride concentration in mg L^{-1}: 1-100, 2- 200, 3-500), 20°C, scan rate 20 mV s^{-1}, rotation rate 500 rpm, reference electrode: Mercury Oxide Electrode at 0.210 V (SHE), IR compensation).

Figure 5 shows the polarization curve obtained by using a boron doped diamond electrode in synthetic drinking water. It is obvious that high electrode potentials are reached whereby many radicals and consecutive products with high oxidation potential can be produced.

As can be seen in Figure 5 higher chloride concentration supports higher current density at the same potential. One reason for this behavior is the acceleration of migration because at low ionic strength ion transport to an electrode may be limited both by diffusion and migration [Bergmann and Koparal, 2005].

Species Generation and Influence Factors

Chloride, active chlorine, chlorate and perchlorate were analyzed in NaCl electrolysis using a rotating MO anode and MO cathode (Figure 6). The experiment is the simplest setup because only chloride is electrolysed. Chloride concentration (with starting value at the limit allowed for drinking water) steadily falls. Active chlorine, whose concentration forms a maximum vs. specific charge flow, is the main product. The chlorate concentration increases linearly. Perchlorate formation is relatively slow with current efficiencies lower than 10% in the starting period. It can be seen from the graph that more perchlorate was found when active chlorine concentration curve had passed the maximum. The behavior shows an important fact – the distribution of charge between several reacting species forms various products in different reactions (active chlorine, chlorate, perchlorate, oxygen etc.). It can be concluded that

- the processes forming active chlorine and chlorate are preferred compared to perchlorate formation,
- the mechanism of chlorate and perchlorate formation is obviously of electrochemical nature. This follows from the linear character of chlorate accumulation and the formation rate of perchlorate which depends on the charge distribution. There are mechanisms for the chemical formation of chlorate too (Eqs. 10-14). But the corresponding overall reaction rate is too low ($k = 10^{-3}$ M^{-1} s^{-1}) to fit the data of the experiments.

Results and Discussion

$$OCl^- + 2HOCl \rightarrow ClO_3^- + 2H^+ + 2Cl^- \quad (10)$$

with the reaction steps

$$2OCl^- \rightarrow ClO_2^- + Cl^- \quad (11)$$

$$OCl^- + ClO_2^- \rightarrow ClO_3^- + Cl^- \quad (12)$$

or as suggested by [Czarnetzki and Jansen, 1992]

$$2HClO \rightarrow ClO_2^- + 2H^+ + Cl^- \quad (13)$$

$$ClO_2^- + HClO \rightarrow ClO_3^- + H^+ + Cl^- \quad (14)$$

If even mixed oxide electrodes with adsorbed reacting species are characterized by charge transfer competition, a more explicit competition should be expected for BDD anodes which adsorb species only weakly. Figure 6 demonstrates clearly the perchlorate formation potential at MO anodes. But it has to be mentioned that the chosen current density of 500 A m^{-2} is relatively high. Typically, technical cells work between 50 and 200 A m^{-2} and at lower specific charge flow. Perchlorate can only rarely be found at remarkable concentration (please see section 4.2 of this chapter).

In analogy to Figure 6 the species behavior in an electrochemical treatment experiment using a BDD anode is shown in Figure 7. Both active chlorine and chlorate concentration curve have maxima. Active chlorine oxidation seems to be very fast and therefore only small concentration values can be found. When the maxima are passed, exponential, i.e. accelerated perchlorate formation can be observed because more charge is available for chlorate oxidation. The Cl balance is in the range of experimental errors (3-5%). This means, probably no other products are present.

The results of the perchlorate production in model water comparing 3 electrode materials are listed in Table 2. Data for platinum is given in [Bergmann et al, 2009b].

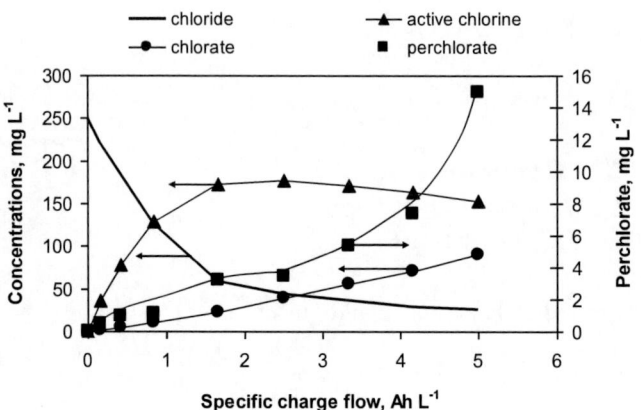

Figure 6. Concentrations versus specific charge flow in a synthetic drinking water system containing chloride (undivided cell with rotating anode, rotation rate 300 rpm, IrO_2/RuO_2 cathode, starting chloride concentration 250 mg L^{-1} (7 mM), current density 500 A m^{-2}, volume 0.1 L, temperature 20°C).

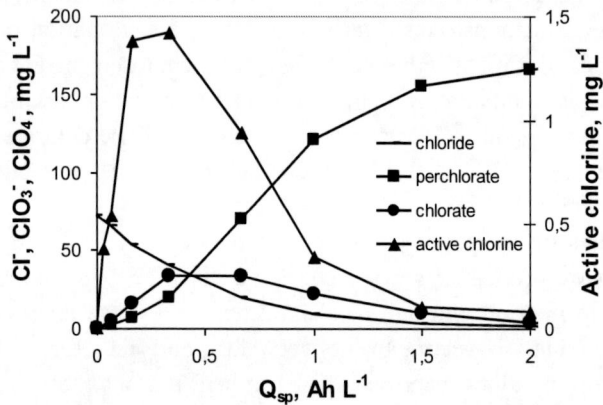

Figure 7. Concentration versus specific charge flow in a synthetic drinking water system containing chloride (undivided cell with rotating BDD anode, rotation rate 300 rpm, IrO_2/RuO_2 cathode, starting chloride concentration 73 mg L^{-1} (2 mM), current density 200 A m^{-2}, volume 0.1 L, temperature 20°C).

Figure 8 illustrates chlorate and perchlorate formation in a semi-technical cell according to the experimental setup shown in Figure 2d. As the cell contains 7 pairs of electrodes, relatively high specific charge flow can be set even in single pass operation. In order to gain a better understanding: at 200 A m^{-2} the specific charge flow amounts 0.2 Ah L^{-1}. Electrolysis results in measurable chlorate and perchlorate concentration which are, however, 2-3 orders of magnitude lower when compared with BDD.

Table 2. Perchlorate concentration in µg L^{-1} (ppb) after 5 minutes of electrolyzing 0.1 L of model water using different rotating anodes (chloride concentration 72-84 mg L^{-1} (2.04-2.36mM), temperature 20°C, anode rotation rate 300 rpm, IrO$_2$/RuO$_2$ cathode)

Current density, A m^{-2}	BDD	IrO$_2$/RuO$_2$	IrO$_2$
100	790	15	8
300	1329	130	60

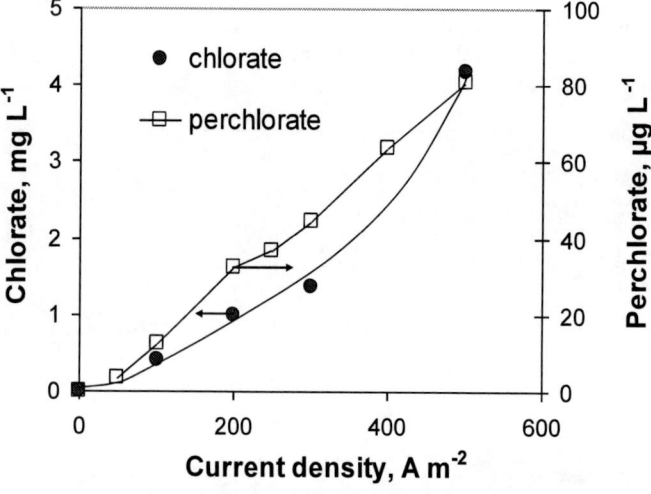

Figure 8. Chlorate and perchlorate formation in a continuous semi-industrial cell using drinking water with chloride addition and varying current density (MO anodes and cathodes, chloride concentration 210 mg L^{-1}, temperature 18.5 °C, flow rate 100 L h^{-1}).

The relationship between the chloride concentration and the perchlorate formation can be seen in Figure 9. At higher initial chloride concentrations less perchlorate is formed because other species consume the charge.

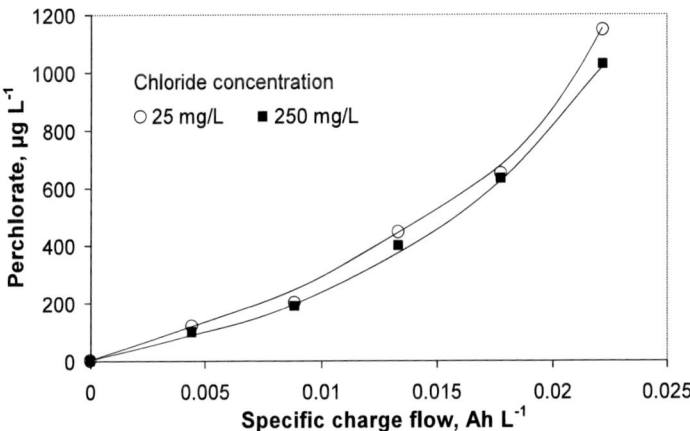

Figure 9. Perchlorate concentration versus specific charge flow in a synthetic drinking water system varying chloride concentration (undivided cell according Figure 2b, BDD anode and cathode, current density 200 A m^{-2}, volume 0.75 L, temperature 20°C).

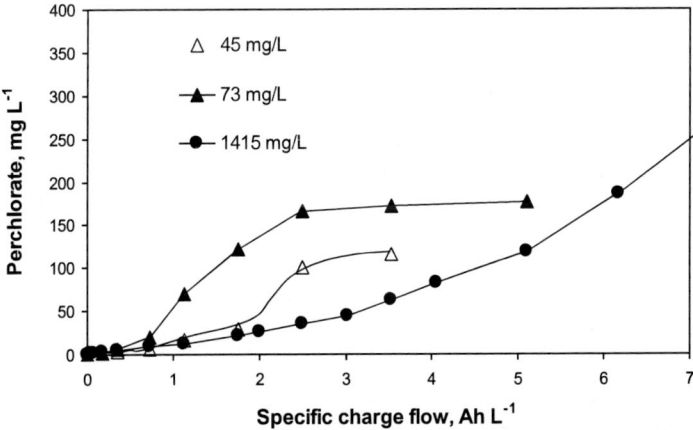

Figure 10. Perchlorate concentration versus specific charge flow in a synthetic drinking water system with varying chloride concentration (discontinuous cell with rotating BDD anode and IrO$_2$/RuO$_2$ cathode, rotation rate 300 rpm, current density 200 A m^{-2}, volume 0.1 L, temperature 20°C).

In Figure 10 perchlorate formation is plotted against the specific charge flow for three different initial chloride concentrations. With higher charge flow or reaction time (it is not shown in Figure 9) perchlorate formation follows the chloride concentration.

The perchlorate formation potential depends on different factors:

- Higher chloride concentrations can cause a higher ion transport to the anode surface enlarging the reaction probability of chloride ions. Therefore, the perchlorate formation may be suppressed due to reaction competitions (Figures 9 and 10). However, more perchlorate is expected over a longer time in discontinuous mode because finally all chloride can be converted to perchlorate if re-reduction effects on the cathode are neglected.
- Higher electrode potential or higher current density favors chlorate and perchlorate formation. Consequently, current densities higher than 150 A m^{-2} should be avoided for higher specific charge flow (long-term or recirculation electrolysis, high electrode surface per volume of flow-through reactors etc.). Sharp electrode edges should also be avoided. Current densities in these places may be multifold higher compared to the average current density [Kodym et al, 2005 and 2006].

Higher flow rates and turbulences accelerate the species supply to the electrode surface but also the transport of intermediates from the electrode surface. As a result specific reactions can be preferred. Kraft et al. reported [Kraft et al., 2006] that in solutions almost free of chloride enhanced ozone production was obtained at higher flow rates. A general assessment of the convection influence on perchlorate formation cannot be made. Moreover, many cells suffer from uncontrolled flow regimes inside the cell [Polcaro et al., 2008] which makes interpretations difficult.

Other influence factors such as temperature and mass transfer have not been studied in detail yet.

BY-PRODUCT FORMATION IN TECHNICAL CELLS EQUIPPED WITH MO ELECTRODES

The generators discussed in this chapter have one in common: the electrolysis of brine by mixed oxide electrodes. On the other hand, the specification of the electrolysis chambers and also their operating conditions are varying. Furthermore, the generators can be subdivided in those producing chlorine gas and those producing hypochlorite solution. In order to understand the relevance of perchlorate formation in the context of drinking water disinfection the whole spectrum of inorganic by-products formed during the electrolysis is discussed in this chapter. Besides perchlorate the most relevant by-products of electrolysis are chlorite, chlorate and bromate. The measured values of the XO_n^- ions (X = Cl, Br; n = 1 - 4) after electrolysis of brine with several generators are summarized in Table. 3.

The concentration of chlorite was below the limit of determination in either case. However, chlorate, perchlorate and bromate were found in the produced hypochlorite stock solution. The level of chlorate formed during electrolysis ranged between a few milligrams per litre to some grams per litre. In contrast, the concentration of perchlorate and bromate was considerable lower. Perchlorate was found in stock solutions of three generators; and although the generator-specific operating conditions were comparable, the concentration of the formed perchlorate varied significantly. For instance, generator A produced between 0.7 and 8.8 mg L^{-1}, generator B 0.26 to 9 mg L^{-1} and generator C less than 0.05 to 2.04 mg L^{-1}.

Bromate was found in all tested stock solutions. But the resulting bromate concentration did not exceed 3.5 mg L^{-1} except for one stock solution (A2).

Additionally, the chlorine specific concentration of the mentioned ions [µg mg^{-1}] was calculated and is listed in Table 3.

The results show that

a) the formation of perchlorate as by-product of technical electrolysis is in principle possible and
b) this process has to be considered in connection with at least the formation of chlorate and probably bromate if the chlorine is used for drinking water disinfection.

Also the results have to be checked considering national and international guidelines.

If the generated stock solution of hypochlorite is used for disinfection, the level of disinfection by-products measured in drinking water is composed of the sum of those which were added to the water with the chlorine (1) and those formed in the water with increasing contact time with chlorine (2).

Results and Discussion 23

In the past, the formation of chlorite, chlorate and bromate as by-products of the disinfection process was examined in more detail in many countries [Rook, 1974, Amy, 1987, Means, 1993, WHO, 2008, EPA, 2008, Hrudey, 2009]. Based on the knowledge of the formation of disinfection and electrolysis by-products and the results of this generator test the relevance of such by-products for drinking water quality can be assessed.

Chlorite: Chlorite is formed when chlorine dioxide is used for water disinfection. The drinking water regulation of the WHO does not define a maximum dosage of the agent. Nevertheless, when chlorine dioxide is used as final disinfectant the resulting chlorite concentration should not exceed 0.2 mg L^{-1}. Therefore, it has to be considered that up to 70 % of the chlorine dioxide will be transferred into chlorite. In principle, chlorite is not found in chlorine electrolysis. So it can be concluded that the dosage of chlorine dioxide is the only relevant source of the formation of chlorite.

Chlorate: Electrochemically generated hypochlorite stock solutions contain chlorate depending on their age and storage conditions [Gordon et al., 1993 and 1995]. Therefore, the chlorate concentration in water can increase according to the chlorine age and dose used for disinfection. The results of the systematic investigations show that between 18 and 275 µg chlorate per mg active chlorine is added to the water. Furthermore, chlorate can also be formed as a disinfection by-product in water depending on the dose of chlorine or chlorine dioxide and the contact time. Recently done systematic examinations show that the formation of chlorate in drinking water does not exceed 50 µg L^{-1} under practical conditions [Schmidt et al., 1999].

Perchlorate: The formation of perchlorate as a disinfection by-product after chlorination cannot be expected. On the other hand, the formation of perchlorate as a by-product of the chlorine generation seems to play a decisive role. The results of the sampling campaign show that between 5 and 92 µg perchlorate per mg active chlorine were brought into the water (Table 3). However, a post formation of perchlorate in the hypochlorite stock solution with increasing age was not observed.

Table 3. Results of by-products formed during electrolysis and input in drinking water per mg chlorine

	ClO_3^- (stock solution) [mg L^{-1}]	ClO_3^- per mg chlorine [µg]	ClO_4^- (stock solution) [mg L^{-1}]	ClO_4^- per mg chlorine [µg]	BrO_3^- (stock solution) [mg L^{-1}]	BrO_3^- per mg chlorine [µg]
A1	2200.0	78	0.69	< 1	3.0	< 1
A2	2300.0	91	0.79	< 1	11.0	< 1
A3	2981.0	142	8.80	< 1	3.5	< 1
B1	27.0	268	9.00	89	< 1.0	< 1
B2	19.0	273	6.10	91	< 1.0	< 1
B3	22.0	72	0.36	< 1	2,0	6
C1	22.0	86	< 0.05	< 1	1.0	4
C2	29.4	71	< 0.05	< 1	0.4	< 1
C3	79.9	209	2.04	5	1.5	4
D1	15.0	82	< 0.05	< 1	1.5	8
D2	3.4	18	< 0.05	< 1	1.4	7
E1	25.6	19	< 0.05	< 1	2.1	2
E2	35.2	22	< 0.05	1	2.2	< 1
E3	34.2	33	< 0.05	1	2.9	3

Bromate: The formation of bromate as a disinfection by-product was well examined in the past [von Gunten and Hoigné, 1992, AWWA, 1993]. Bromate is formed when ozone is used for disinfection of bromide containing water. Electrochemical bromate formation in drinking water matrices was first mentioned in [Kresse et al., 2008, Bergmann et al. 2008, Bergmann et al., 2009c]. Results of examinations done by TZW Dresden show that a content of bromate up to 10 µg per mg active chlorine is possible in electrochemically generated active chlorine solutions. A post formation of bromate in the stock solution and its formation in drinking water after chlorine disinfection were not observed.

All results show that compared to chlorite the formation of chlorate, perchlorate and bromate during chlorine electrolysis seems to be a relevant process, which can influence the quality of the chlorine stock solution and for this reason the whole water disinfection process in a decisive way. So far, the chief causes which influence the level of such a halogen oxyanion formation during chlorine electrolysis from brine are not well understood due to their complexity. Essential factors seem to be the size of the cell, the quality of the brine, especially the bromide content, the material of the electrodes and the operating conditions of the electrolysis. Therefore, a first comparison of all tested generators with respect to the operation parameters can help to get a first impression of the general situation during chlorine electrolysis. A first approach to explain the mechanisms is given in the next section of this chapter.

Table 4. Percentage of the conversion of chloride and bromide into hypochlorite, chlorate, perchlorate and bromate during electrolysis

	$HOCl/OCl^-$ [%]	ClO_3^- [%]	ClO_4^- [%]	BrO_3^- [%]
A1	36	1.7	4.5E-04	< 22
A2	32	1.8	5.2E-04	< 69
A3	27	2.3	5.8E-03	~100
B1	0.4	0.1	1.9E-02	-
B2	0.3	0.1	1.3E-02	-
B3	1.2	0.1	7.6E-04	< 100
C1	2.8	0.2	-	< 62
C2	4.6	0.2	2.3E-03	< 25
C3	4.2	0.5	1.2E-02	< 94
D1	0.1	3.2E-03	-	< 94
D2	0.1	7.7E-04	-	< 88
E1	1.6	0.02	-	29
E2	1.9	0.03	-	~100
E3	1.3	0.03	-	~100

The percentage of chloride as well as bromide transformation from oxidation state -1 into +1 (hypochlorite), +5 (chlorate, bromate) and +7 (perchlorate) is shown in Table 4. In general, it seems to be clear that the chlorine transformation to +5 is significant higher than to +7. The bromide of the brine is almost completely converted to bromate in many cases. The percentage ranges between < 25 and 100.

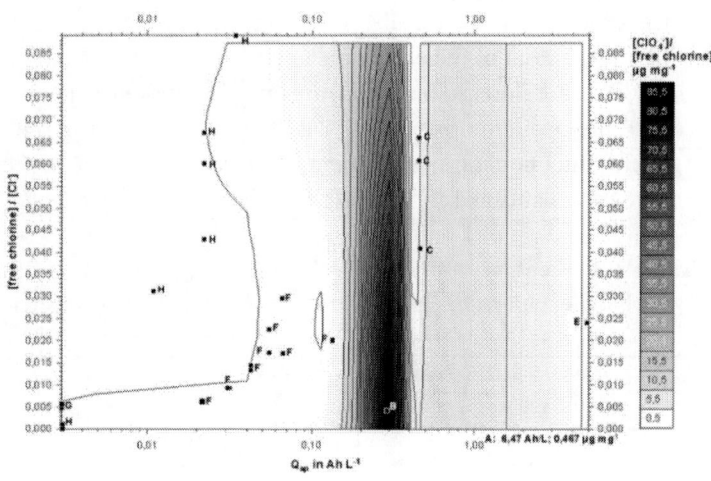

Figure 11. Chlorine- and charge-specific formation of perchlorate for different generators.

Figure 11 depicts the chlorine-specific formation of perchlorate [µg mg^{-1} chlorine]. The specific perchlorate input per mg active chlorine c_{sp} [µg mg^{-1}] is correlated to the effective chlorine generation Cl_{eff} (chlorine/chloride – ratio [mg mg^{-1}]) and the specific charge flow Q_{sp}. The diagram shows the wide range of operating conditions of all tested generators. However, the impact of cell size and electrode material cannot be considered in this way. It can clearly be seen that the perchlorate input into drinking water of most of the tested devices is very low (< 1 µg L^{-1}). The risk seems to increase if the actual efficiency of chlorine generation is comparatively low. But a significant impact of Q_{sp} cannot be seen. From these results it can be concluded that the maximum of perchlorate formation at Q_{sp} = 0.29 Ah L^{-1} is in particular influenced by the electrode material and the construction of the electrolysis cell which both are issues belonging to the top secrets of the companies. It is remarkable that the profile of chlorate and bromate formation is nearly the same as that of perchlorate formation shown in Figure 11.

Chapter 5

DISCUSSION OF PROBABLE MECHANISMS

GENERAL CONSIDERATION

When chloride is converted to perchlorate its oxidation number changes by 8.

$$Cl^- \xrightarrow{H_2O;\ 8e^-} ClO_4^- \tag{15}$$

An electrochemical process with a parallel transfer of eight electrons does not exist. Therefore, the perchlorate formation has to proceed by a stepwise electron transfer via different intermediates. Because hypochlorite and chlorate are always present during electrolysis it is usually assumed that both substances are precursors for perchlorate. Some researchers even report the occurrence of short-lived chlorite at MO and BDD anodes [Bergmann et al., 2009e, Polcaro et al., 2008]. It has to be emphasized that a variety of mechanisms of perchlorate and intermediate formation is probable - due to the high complexity of the systems and operating conditions. In addition different reaction zones can be distinguished:

- *Electrode surface with species in adsorbed state.* The residence time of the species at the surface depends on bonding strength, diffusion direction and rate, surface concentration and electron density and other influences on the reaction rate.
- *Reaction layer inside the diffusion layer near the electrode.*

- *Bulk of the solution.* Reactions in this zone have a chemical character and may continue after switching off the electrolysis. Although it is quite big, this zone mostly contributes to conversion to a relatively low extent. Nevertheless, in the literature, most researchers calculate actual obtained reaction constants with respect to the bulk concentration.

So all typical phenomena (mass transfer by different mechanisms, adsorption, desorption, hydration, dehydration, electron transfer etc.) have to be taken into account to explain reaction mechanisms. The number of publications concerning electrochemical disinfection with regard to the mechanisms is very low, though.

The rate of an electron transfer reaction

$$A + e^- \rightarrow B \tag{16}$$

and the rate of radical reaction may be comparable. For the latter, rate constants up to 10^{11} M^{-1} s^{-1} can be found experimentally.

Applying Faraday's law the maximal perchlorate formation rate can be estimated:

$$\frac{dn_{PC}}{dt} = \frac{I}{8 \cdot F} \tag{17}$$

It can be expected that reaction rates of this magnitude are a result of very fast electrochemical steps and/or fast radical reactions. In this case, slower chemical reactions are negligible. The assumption seems to be correct because perchlorate current efficiencies can be calculated in the range of a few per cents, whereby the losses can be explained by competing reactions which are difficult to avoid.

POSSIBLE MECHANISMS

The reaction of a chloride ion must be assumed as the very first reaction step in perchlorate formation but even the mechanism of this reaction is not clear. More or less each electrochemical reaction depends on the conditions of the electrode surface. Numerous papers deal with chloride oxidation at several MO electrode materials [Trasatti, 2000]. In general, it must be distinguished between chloride ion oxidation by electron transfer with the anode and chloride ion oxidation by reactions with other species. A series of radicals (hydroxyl radical, nitrate radical and others) belong to this group of species. If the chloride concentration is high enough to cover all active sites on the electrode surface the direct electron transfer reaction is more probable. In contrast, mainly the participation of water can be expected in highly diluted systems (drinking water conditions) at the first stage of electron transfers.

Apart from that, some researches [Ferro et al., 2000], suggest that chloride reacts by direct electron transfer to chlorine radicals at BDD anodes with further chlorine formation:

$$Cl^- \rightarrow Cl^* + e^- \tag{18}$$

$$Cl^* + Cl^* \rightarrow Cl_2 \tag{19}$$

The suggestion is questionable when water of low ionic strength is electrolyzed. The water concentration of a 35 mg [Cl$^-$] L^{-1} solution (1 mM) is more than 10 000 times higher than the chloride concentration [Bergmann, 2010a]. Consequently, water oxidation to hydroxyl radicals and the chloride reaction with the generated hydroxyl radicals are highly probable:

$$Cl^- + {}^*OH \rightarrow Cl^* + OH^- \tag{20}$$

The *OH radical is formed from a hydroxide ion or a water molecule:

$$OH^- \rightarrow {}^*OH + e^- \quad E^0 = 1{,}70\text{-}2{,}02 \text{ V (SHE)} \tag{21}$$

$$H_2O \rightarrow {}^*OH + H^+ + e^- \quad E^0 = 2{,}59\text{-}2{,}85 \text{ V (SHE)} \tag{22}$$

Usually reaction 6 is written in the form of an adsorbed (weakly adsorbed on BDD) hydroxyl radical at the electrode substrate S:

$$S + H_2O \rightarrow S\text{-}OH + H^+ + e^- \tag{23}$$

At sufficient high electrode potential, further oxidation to oxygen is possible:

$$S\text{-}OH \rightarrow S\text{-}O + H^+ + e^- \quad (24)$$

$$2S\text{-}O \rightarrow 2S + O_2 \quad (25)$$

According to Suffredini et al. [Suffredini et al., 2004] reaction 23 is the rate determining step. In experiments it was found that obviously the diffusion of the hydroxyl radicals causes an additional limitation [Bergmann et al., 2008]. Furthermore, a desorption process may exist which also slows down the reaction. As can be seen from Figure 5 and 12 limiting current density plateaus exist in the BDD polarograms when diluted systems are studied. In contrast to the second plateau the first plateau does neither depend on the concentration nor on the rotation rate. It is always in the range between 10 and 20 A m^{-2}. The curve progression can be explained by a possible limited surface diffusion of the *OH radicals, which can to a low extent react to hydrogen peroxide or oxychlorine compounds. The second plateau can be caused by the reaction of the species with radicals of a higher oxidation state (O*).

Dissolved chlorine reacts with water and faster with hydroxide ions if the pH is higher than approximately 3 [White, 1999]:

$$Cl_2 + OH^- \rightarrow OCl^- + H^+ + Cl^- \quad k = 5 \times 10^{14} \quad (26)$$

$$Cl_2 + H_2O \leftrightarrow HOCl + H^+ + Cl^- \quad pK = 3.3 \quad (27)$$

$$HOCl \leftrightarrow H^+ + OCl^- \quad pK = 7.4 \quad (28)$$

Reactions 26-28 immediately limit the formation of further oxychlorine products at current efficiencies lower than 50%. In fact, all experiments resulted in current efficiencies less than 50 % for active chlorine and chlorate formation, whereas chloride depletion (at least for a short time in the beginning of the experiments) reached nearly 100%). During electrolysis of chlorate to perchlorate, current efficiencies higher than 50% were measured. Similar to the chemical formation of chlorate and bromate in advanced oxidation processes [v. Gunten, 2003] the OX$^-$ form seems to play a decisive role.

Discussion of Probable Mechanisms 31

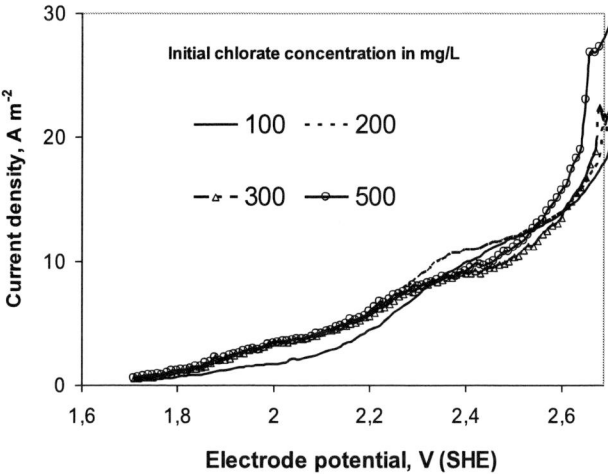

Figure 12. Polarization curves on a rotating BDD anode in NaClO$_3$ solution varying chlorate concentration (20°C, scan rate 20 mV s^{-1}, 500 rpm, reference electrode: Mercury Oxide Electrode at 0.210 V (SHE), IR compensation).

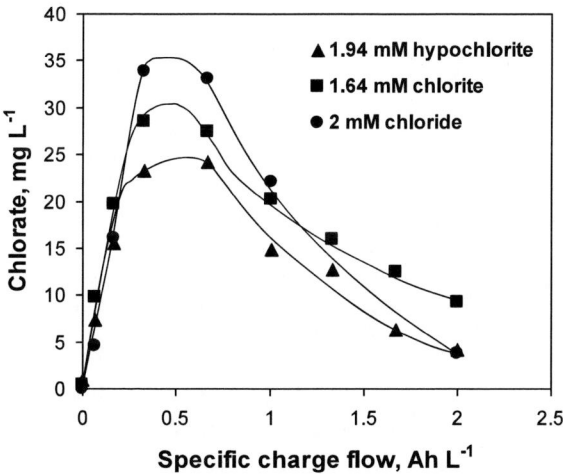

Figure 13. Perchlorate concentration versus specific charge flow in electrolysis experiments using different probable perchlorate formation intermediates as initial substances (discontinuous cell with rotating BDD anode and IrO$_2$/RuO$_2$ cathode, rotation rate 300 rpm, current density 200 A m^{-2}, volume 0.1 L, temperature 20°C).

Figure 13 shows the chlorate formation at BDD from different species containing Cl of comparable starting concentration and at 200 A m⁻². There are no differences at the beginning. The corresponding maximum current efficiency amounts to 22.5 % for the chloride, 15 % for the hypochlorite, and 7.5 % for the chlorite experiment. Differences in chlorate formation efficiency become visible at higher specific charge flow. Despite the longest oxidation path the current efficiency of the chloride is highest. This can be explained by differences in the bulk concentration and concentration influence of re-reduction processes at the cathode.

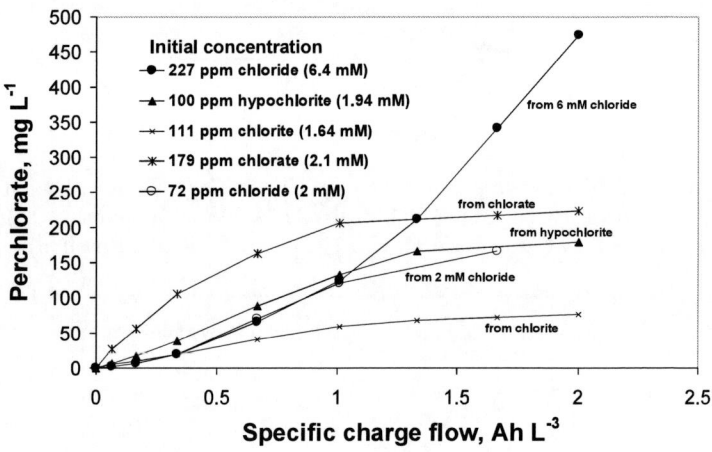

Figure 14. Perchlorate concentration versus specific charge flow in electrolysis experiments using different probable perchlorate formation intermediates as initial substances (discontinuous cell with rotating BDD anode and IrO_2/RuO_2 cathode, rotation rate 300 rpm, current density 200 A m⁻², volume 0.1 L, temperature 20°C).

Both hypochlorite and chlorite can be reduced in cathodic reactions. But temporally formed intermediates of low concentration rather remain near the anodic surface where they are oxidized instead of being reduced at the cathode.

Figure 14 presents the resulting perchlorate concentration profiles vs. the specific charge flow for different initial substances. In addition, experiments with 6.4 mM chloride solution and 2.2 mM chlorate electrolyte can be seen. As expected, the latter reveals the highest perchlorate formation rate. The curve for the 6.4 mM chloride solution confirms the discussion visualized in Figures 9 and 10.

Discussion of Probable Mechanisms

A maximum current efficiency of 26 % can be calculated from the first measuring point for the oxidation of chlorate to perchlorate. A maximum current efficiency of 32,5 % was calculated for the oxidation of chloride (2 mM), 21 % for the oxidation of hypochlorite and 9 % for the oxidation of chlorite to perchlorate. Most values exceed the maximum values obtained from limiting current densities as shown in Figure 12. The simplest explanation is the participation of O* radicals in reactions of the following kind:

$$X^- + O^* \rightarrow XO^- \tag{29}$$

or

$$XO_n^- + O^* \rightarrow XO_{n+1}^- \tag{30}$$

The participation of O* radical is accepted for the mechanism described in equation 7 for several electrode materials and in ozone formation on BDD:

$$O_2 + O^* \rightarrow O_3 \tag{31}$$

Djordevic and co-workers [Djordevic et al., 1973] conclude from hypochlorite polarization experiments new reaction paths for chlorate on platinum in the potential region for oxygen evolution. Cettou and co-workers make a suggestion regarding the bromate formation from bromide on RuO_2 anode [Cettou et al., 1984]:

$$OBr^- + 2O^* \rightarrow BrO_3^- \tag{32}$$

Buxton and Subhani confirm oxidation reactions of triplet oxygen atoms with hypochlorite to chlorite and with chlorite to chlorate [Buxton and Subhani, 1972].

Janssen and van der Heyden [Janssen and v.d. Heyden, 1995] exclude this mechanism for perchlorate formation from chlorate on a platinum anode, because with increasing potential no decreasing oxygen formation rate was observed. They suggest the following electrochemical mechanism:

$$ClO_3^- + OH_{ads} \rightarrow ClO_4^- + H^+ + e^- \quad (33)$$

(Remark: This reaction is not possible in one step as often written). Munichandraiah and Aathyanarayana [Munichandraiah and Aathyanarayana, 1987] found that ionic strength seems to have an influence on kinetics. In addition to the mentioned mechanism they propose a further one for the perchlorate formation from chlorate on β-PbO_2 (Ti) anodes via a chlorate radical:

$$ClO_3^- \rightarrow (ClO_3)_{ads} + e^- \quad (34)$$

$$(ClO_3)_{ads} + OH_{ads} \rightarrow ClO_4^- + H^+ + e^- \quad (35)$$

In the literature all effects are often related to *OH radicals because the formation of these radicals at BDD anodes is approved in most instances by scientist. In our opinion, this assumption can only be applied for relatively high surface concentrations of the reacting species (for instance organic molecules in excess), [Bergmann, 2010b]. If species react under limiting conditions, other mechanisms should be included into consideration. Nevertheless, under drinking water conditions *OH radicals do contribute to a certain extent to the reaction results. The possible formation of Cl*, OCl*, ClO_2, ClO_3^* from Cl⁻ or ClO_n^- species and *OH, and even ClO_4* formation are known from radiolysis experiments. Thus, the simplest scheme would include (starting consideration from OCl⁻) reactions of the chemical type such as

$$*OH + Cl^- \leftrightarrow HOCl^- \quad (36)$$

with further reactions of the activated complex HOCl⁻, and

$$ClO_n^- + *OH \rightarrow ClO^*_{n+1} + H^+ \quad (37)$$

Von Gunten [v. Gunten, 2003] estimates a reaction rate in the range of 10^5 M^{-1} s^{-1} at pH 5 for the overall reaction of hydroxyl radicals with chloride ions forming chlorine radicals with the intermediate HOCl⁻ (Eq. 36).

Table 5 summarizes the reaction constants for *OH reactions with several species.

Table 5. Kinetic constants reported in literature for reactions of the type Y + *OH→

Y:	Cl⁻	OCl⁻	HOCl	ClO_2^-	ClO_2	ClO_3^-	*OH
k, $M^{-1}s^{-1}$	3-4.3x10^9	9x10^9	0.11-0.14 x10^9	6.1-7.9x10^9	1.4-4.4 x10^9	< 1x 10^6	1-2.8x10^{10}

Specific reactions can result in species of lower oxidation state than the starting oxychlorine compound. It is notable, that a very high reaction rate constant (last column) is known for the hydrogen peroxide formation:

$$\text{*OH} + \text{*OH} \rightarrow H_2O_2 \qquad k = 1 - 2.8 \times 10^{10} \text{ M}^{-1} \text{ s}^{-1} \qquad (38)$$

Reactions of the type described in Eq. 39 for species in an adsorbed state would have an electrochemical character.

$$ClO^*_n + \text{*OH} \rightarrow ClO_{n+1}^- + H^+ + e^- \qquad (39)$$

But this assumption is questionable and cannot account for an explanation of the limiting current density for *OH radicals.

Probably, intermediates such as Cl_2, Cl_2O_2, and Cl_2O_6 are formed from the corresponding radicals. Cl_2 is known as dissolved chlorine. Emmenegger and Gordon [Emmenegger and Gordon, 1967] considered Cl_2O_2 as an important intermediate in chlorate formation for the first time:

$$Cl_2 + ClO_2^- \rightarrow Cl_2O_2 + Cl^- \qquad (40)$$

It may react to chlorine dioxide or with water to chlorate:

$$Cl_2O_2 + H_2O \rightarrow ClO_3^- + Cl^- + 2H^+ \qquad k \text{ ca. } 10^3 \text{ M}^{-1} \text{ s}^{-1} \qquad (41)$$

Hydrolysis of ClO_3 or Cl_2O_6 gives chloric and perchloric acid:

$$Cl_2O_6 + H_2O \rightarrow HClO_3 + HClO_4 \qquad (42)$$

Hydrolysis of ClO_3 and Cl_2O_6 gases also results in perchlorate formation.

Chloric acid may disproportionate:

$$2HClO_3 \leftrightarrow HClO_4 + HClO_2 \qquad (43)$$

Though theoretically possible, the mentioned reactions only are probable if the rate determining reactions have reaction constants in the range of those for fast radical reactions (10^8-10^{10} M^{-1} s^{-1}).

As already mentioned above *OH radicals are not exclusively generated at BDD anodes. Bergmann et al. [Bergmann et al., 2008] found during electrolysis with RNO comparable *OH concentrations when IrO_2, IrO_2/RuO_2 and BDD were used as anode material in chloride-free water. Similar effects were reported by Jeong and co-workers [Jeong et al., 2009].

Certain amounts of chlorate and perchlorate can be generated by applying ozone. In experiments performed by Kraft et al [Kraft et al., 2006] ozone current efficiencies of 5% were estimated for drinking waters of very low chloride concentration at 200 A m^{-2}.

It can be expected that the influence of ozone-based reactions increases if higher current densities are applied. Kinetic studies of ozonation were published by von Gunten [v. Gunten, 2003]. For example, chlorine dioxide quickly reacts to chlorate or ClO_3*/Cl_2O_6. Different formula expressions are given in the literature:

$$2ClO_2 + O_3 + H_2O \rightarrow 2ClO_3^- + O_2 + 2H \qquad (44)$$

$$ClO_2 + O_3 \rightarrow ClO_3^- + O_2 \qquad (45)$$

$$2ClO_2 + O_3 \rightarrow Cl_2O_6 + 0.5O_2 \qquad (46)$$

Second order reaction constants in the range of 10^3 M^{-1} s^{-1} can be found for these reactions. Hydrolysis results in the formation of perchloric acid (Eqs. 41-42). Chlorine dioxide can be formed from chlorite and ozone (k = 8.2 x 10^6 M^{-1} s^{-1}). Ozone does not react with hypochlorous acid. It reacts slowly with hypochlorite.

Ozone is able to react with hydroxyl radicals if no other reaction partners are available:

$$*OH + O_3 \rightarrow HO_2^* + O_2 \quad\quad k= 3 \times 10^9 \text{ M}^{-1}\text{ s}^{-1} \quad\quad (47)$$

Reactions with chlorite ($k=4\times10^6$ M^{-1} s^{-1}), hypochlorite ($k=10^3$M^{-1}s^{-1}), chlorate ($k < 10^{-4}$ M^{-1} s^{-1}) or deprotonated hydrogen peroxide are slower.

Many other reactions such as oxidation reactions with participation of Cl or NO$_3$ radicals are still possible. But these reactions are not discussed here because the occurrence of Cl* or *NO$_3$ radicals is not proven yet and therefore all discussions about them are speculative.

Chapter 6

CONCLUSIONS

- Discussing the electrochemical perchlorate formation in diluted aqueous systems a complex and confusing picture is obtained due to the high complexity of the reaction conditions. Therefore, in practical application no general assessment and statements can be given without preliminary experiments and detailed knowledge of process parameters
- Despite the complexity some main parameters help to confine the perchlorate formation conditions. Low ionic strength of the solution and high electrode potentials belong to these conditions. High amounts of specific charge flow through the system are another important reason for perchlorate accumulation. Indirectly, factors such as high applied current density, geometry favoring high local current density or special electrode materials characterized by enlarged electrode potential (BDD) have significant influence on perchlorate formation. The role of hydrodynamics and temperature has not yet been studied sufficiently.
- After brine electrolysis perchlorate was found in the resulting hypochlorite stock solution. But the input into drinking water through disinfection is in most cases lower than 10 $\mu g\ L^{-1}$ because of the dilution when given into the water.
- A higher chloride concentration does not necessarily increase perchlorate formation. If electrical charge is easier consumed in competing reactions, even lower perchlorate concentrations can

be measured for certain time intervals. However, if the electrochemical treatment is extended to very long times, all chloride is converted to perchlorate except the amount of re-reduced products at the cathode in an undivided electrolyser.
- The chemistry of perchlorate formation is obviously related to the generated radicals. The role of O* radicals is still unclear but there are indications that not only *OH radicals take part in the different reactions. From theoretical considerations it can be concluded that the main reaction zone is less than 1 µm near the anode surface because kinetic constants are very high for possible reactions.
- All in all it can be concluded that perchlorate may be formed at many different anode materials. BDD anodes have a higher risk of perchlorate formation than other materials. It is possible that the recommended limiting value for perchlorate in the USA is exceeded. Therefore, it is suggested to set up certification routines for cell producers and technology users. First approaches are known from German projects [Bergmann, 2009f].

REFERENCES

AWWA (1993). Brominated DBPs [editorial]. *J. AWWA*, *85*, 41.
Amy, G. L., Chadick, P. A. & Chowdhury, Z. K. (1987). Developing Models for Predicting Trihalomethane Formation Potential and Kinetics. *J. AWWA.*, *79*, 89-97.
Bergmann, H., Iourtchouk, T., Schoeps, K. & Ehrig, F. (2001). What is the so-called Anodic Oxidation and what can it do? (in Germ.). *GWF Wasser Abwasser*, *142*, 856-869.
Bergmann, M. E. H. & Koparal, A. S. (2005). Studies on electrochemical disinfectants production using anodes containing RuO_2, *J. Appl. Electrochem.*, *35*, 1321-1329 and *Erratum*, 2006, 36, 845-846.
Bergmann, H. (2006). On the application of DPD method for disinfection drinking water electrolysis, Germ.: *Zur Anwendung der DPD-Methode bei der Desinfektionselektrolyse von Trinkwasser,* gwf Wasser-Abwasser, *147*, 780-786.
Bergmann, M. E. H. & Rollin, J. (2007a). Product and by-product formation in disinfection electrolysis of drinking water using boron-doped diamond anodes. *Catalysis Today*, *124*, 198-203.
Bergmann, H., Rollin, J., Czichos, C. & Roemer, D. (2007b). Perchlorate analysis in drinking water electrolysis-a new application for Ion Chromatography (in Germ.). *Labo, 38*, 26-28.
Bergmann, H., Iourtchouk, T., Borutzky, U. & Kodym, R. (2008). *Erweiterte Grundlagenuntersuchungen zur Entwicklung mobiler Wasserversorgungssysteme mit erhöhter Desinfektionswirkung*, Published Research Report, Anhalt University, Köthen FKZ 1721X04.

Bergmann, M. E. H. (2009a). Drinking Water Disinfection by In-line Electrolysis-Product and Inorganic By-Product Formation. In: Comninellis, Ch., Chen, G. (eds.) *Electrochemistry for the Environment*, Springer, 163-205.

Bergmann, M. E. H., Rollin, J. & Iourtchouk, T. (2009b). The occurrence of perchlorate during drinking water electrolysis using BDD electrodes. *Electrochim. Acta, 54*, 2102-2107.

Bergmann, H., Iourtchouk, T. & Rollin, J. (2009c). *German patent application*, 2009, 10, 040 651. 4.

Bergmann, H., Iourtchouk, T., Rollin, J. & Kresse, C. (2009d). *The Occurrence of Perbromate on BDD during Water Electrolysis in ppm Range of Bromide Concentration*, poster, Book of Abstracts, Annual ISE Meeting, Beijing (China).

Bergmann, M. E. H., Schmidt, W. & Dommaschk, A. K. (2009e). *Detection of ClO_2 and ClO_3^- in electrolysed water of very low ionic strength using LGB method and IC*, poster, Annual Meeting of the International Electrochemical Society, Book of Abstracts, Annual Beijing (China).

Bergmann, M. E. H. (2009f). *Save Electrochemical Disinfection of Drinking Water*-a New Joint Research Project in Germany, poster, Book of Abstracts, Annual Meeting of the International Electrochemical Society, Beijing (China).

Bergmann, H. & Iourtchouk, T. (2010a). *On halogenate and perhalogenate formation using boron doped diamond electrodes-state of knowledge and relevance for protection of environment* (in Germ.), Jahrbuch Oberflächentechnik, Bd. 66, Leuze Verlag, Saulgau (in press)., Leuze Verlag, Saulgau, 282-291.

Bergmann, M.E.H. (2010b). *Mechanistic consideration of OH radical behavior on BDD anodes*. Lecture, Annual Meeting of the International Society of Electrochemidstry, Nice/France.

Buxton, G. V. & Subhani, M. S. (1972). Radiation chemistry and photochemistry of oxychlorine ions. Part 2.-Photdecomposition of aqueous solutions of hypochlorite ions. *J. Chem.Soc., 68*, 958-969.

Cettou, P., Robertson, P. M. & Ibl, N. (1984). On the electrolysis of aqueous bromide solutions to bromate. *Electrochim. Acta, 29*, 875-885.

References

Czarnetzki, L. R. & Janssen, L. J. J. (1992). Formation of hypochlorite, chlorate and oxygen during NaCl electrolysis from alkaline solutions at a RuO_2/TiO_2 anode. *J. Appl.Electrochem.*, *22*, 315-324.

D'Ans, J. & Freund, H. E. (1957). Kinetic studies 1. On the formation of chlorate from hypochlorite (in Germ.). *Zeitschrift fuer Elektrochemie*, *61*, 10-18.

Djordević, A. B., Nikolić, B. Ž., Kadija, I. V., Despić, A. R. & Jakšić, M. M. (1973). Kinetics and mechanism of electrochemical oxidation of hypochlorite ions. *Electrochim. Acta*, *18*, 456-471.

DIN EN ISO 15061 (2001). *Wasserbeschaffenheit - Bestimmung von gelöstem Bromat - Verfahren mittels Ionenchromatographie* (ISO 150, 2001): German issue: EN ISO 15061:2001, Deutsches Institut für Normung e.V., Berlin.

Emmenegger, F. & Gordon, G. (1966). The apid interaction between sodium chlorite and dissolved chlorine. *Inorgan. Chem.*, *6*, 633-635.

EPA (2008). *Interime Drinking Water Health Advisory for Perchlorate*. EPA 822-R-08-025.

Ferro, S., De Battisti, A., Duo, I., Comninellis, Ch., Haenni, W. & Perret, A. (2000). Chlorine evolution at highly boron-doped diamond electrodes. *J. Electrochem. Soc.*, *147*, 2614-2619.

Fujishima, A., Einaga, Y., Rao, T. N. & Tryk, D. (eds.) (2005). *Diamond Electrochemistry*. Elsevier, Amsterdam.

Gordon, G., Adam, L. C., Bubnis, B. P. & Wliczak, A. (1993). Controlling the formation of chlorate ion in liquide hypochlorite feedstocks. *J. AWWA*, *85*, 89-97.

Gordon, G., Adam, L. & Bubnis, B. (1995). Minimizing chlorate ion formation. *J. AWWA*, *87*, 97-106.

Gu, B. (2006). *Perchlorate, Environmental Occurrence, Interaction and Treatment*. Springer, N.Y.

Gutknecht, J., Hartmann, F., Kirmaier, N., Reis, A. & Schoeberl, M. (1981). Anodic oxidation as a water disinfecting process in food plants and breweries (in Germ.). GIT *Fachz. Lab.*, *25*, 472-481.

Hrudey, St.E. (2009). Chlorination and disinfection by-products, public health risk tradeoffs and me. *Wat. Res.*, *43*, 2057-2092.

Jackson, A., Arunagiri, S., Tock, R., Anderson, T. & Rainwater, K. (2004). Technical Note: Electrochemical generation of perchlorate in municipal drinking water systems. *Journ. AWWA*, *96*, 103-108.

Janssen, L. J. J. & v.d. Heyden, P. D. L. (1995). Mechanism of anodic oxidation of chlorate to perchlorate on platinum electrodes. *J. Appl. Electrochem.*, *25*, 126-136.

Jeong, J., Kim, C. & Yyoon, J. (2009). The effect of electrode material on the generation of oxidants and microbial inactivation in the electrochemical disinfection processes. *Wat. Res.*, *43*, 895-901.

Jung, Y.J., Baek, K.W., Oh, B.S. & Kang, J. W. (2010). An investigation of the formation of chlorate and perchlorate during electrolysis using Pt/Ti electrodes: The effect of pH and reactive oxygen species and the results of kinetic studies. *Wat. Res.*, 44, 5345-5355.

Martinez-Huitle, C. A. & Brillas, E. (2008). Elektrochemische Alternativen für die Trinkwasserdesinfektion. *Angew. Chem.*, *120*, 2024-2032.

Munichandraiah and Aathyanarayana (1985). Kinetics and mechanism of anodic oxidation of chlorate ion. *J. Appl.Electrochem.*, *17*, 33-48.

Kirk-Othmer (1979). *Encyclopedia of Chemical Technology.* 3.ed., vol. 5, Wiley, N.Y., 646-666.

Kodym, R., Bergmann M. E. H. & Bouzek, K. (*2005*). *First results of modeling geometry factors in electrolysis cells for direct drinking water disinfection.* Poster *5-042*-P, *56*[th] Annual Meeting of the International Society of Electrochemistry - Busan, Korea/ September *26-30*, Book of abstracts, *896*.

Kodym, R., Bergmann, H. & Bouzek, K. (2006). *The study of geometrical factors of the electrolysis cell for drinking water disinfection with respect to chlorine evolution efficicieny.* Poster, 57th Annual Meeting of the International Society of Electrochemistry, 27. Aug.-1.Sept., Edinburgh/UK.

Kraft, A., Stadelmann, M., Wünsche, M. & Blaschke, M. (2006). Electrochemical ozone production using diamond anodes and a solid polymer electrolyte. *Electrochem. Comm.*, *8*, 883-886.

Kresse, K., Rollin, J. & Bergmann, H. (2008). *Anhalt Conference of Young Scientists, Koethen,* Book of papers (ISBN 978-3-86011-022-5), 27-31.

Kuhn, A. T. (1971). *Industrial Electrochemical Processes.* Elsevier, Amsterdam.

Means, E. G. & Krasner, S. W. (1993). D-DBP Regulations: Issues and Ramifications. *J. AWWA*, *85*, 68-73.

References

Polcaro, A. M., Vacca, A., Mascia, M. & Ferrara, F. (2008). Product and by-product formation in electrolysis of dilute chloride solutions. *J. Appl. Electrochem.*, *38*, 979-984.

Reis, A. (1951). The anodic oxidation as an inactivator of pathogenic substances and processes (in Germ.). *Klin. Wschr.*, *29*, 484-485.

Reis, A. & Henninger, T. (1953). Destruction of malignant growth energy by anodic oxidation (in Germ.). *Klin. Wschr.*, *31*, 39-40.

Reis, A. (1976). Sterilization and decomposition of noxious organic substances by anodic oxidation (in Germ.) *GIT Fachz. Lab.*, *20*, 197-204.

Reis, A. (Editor) (1981). *Anodische Oxidation in der Wasser- und Lufthygiene*, GIT Verlag, Darmstadt.

Rook, J. J. (1974). Formation of haloforms during chlorination of natural waters. Water *Treatment and Examination*, *23*, 234-243.

Schmidt, W., Böhme, U., Sacher, F. & Brauch, H. J. (1999). Formation of chlorate in disinfection of drinking water (in Germ.). *Vom Wasser*, *93*, 109-126.

Sellers, K. et al. (2007). *Perchlorate, Environmental Problems and Solutions*, Taylor and Francis, Boca Raton.

Suffredini,H. B., Machado, S. A. S. & Avaca, L. A., J. (2004). The Water Decomposition Reactions on Boron-Doped Diamond Electrodes. *Braz. Chem. Soc.*, *15*, 16-21.

Tasaka, A. & Tojo, T. (1985). Anodic oxidation mechanism of hypochlorite ion on platinum electrode in alkaline solution. *J. Electrochem Soc.*, *132*, 1855-1859.

Tock, R. W., Jackson, W. A., Anderson, T. & Arunagiri, S. (2004). Electrochemical generation of perchlorate ions in chlorinated drinking water. *Corrosion*, *60*, 757-763.

Trasatti, S. (2000). Electrocatalysis: Understanding the success of DSA. *Electrochimica Acta*, *45*, 2377-2385.

Udupa, H. V. K., Narasimham, K. M., Nagalingam, M., Thiagarajan, N., Subramanian, G., Palanisamy, R., Pushpavanam, S., Sadagopalan, M. & Goüalakrishnan, V. (1971). Large-scale preparation of perchlorates directly from sodium chloride. *J. Appl. Electrochem.*, *1*, 207-212.

V. Gunten U. & Hoigné, J. (1992). Factors controlling the formation of bromate during ozonation of bromide-containing waters. *Journal of Water Supply:Research and Technology–Aqua*, *41*, 299–304.

v. Gunten, U. (2003). Ozonation of drinking water: Part II. Disinfection and by-product formation in presence of bromide, iodide or chlorine, *Wat. Res.*, *37*, 1469-1487.

White, G. C. (1999). *Handbook of chlorination and alternative disinfectants*. 4. ed., John Wiley & Sons, N.Y.

WHO (2008). Guidelines for Drinking-water Quality. *Third Edition incorporating the first and second addenda.* Volume *1*, Recommendations, Geneva/Switzerland.

INDEX

#

20th century, 1, 6

A

acid, 1, 6, 36
active chlorine, xi, 16, 17
active site, 14, 29
adsorption, 14, 28
age, 23
aqueous solutions, 42
assessment, 21, 39
atoms, 13, 33
authorities, ix

B

Beijing, 42
bonding, 27
Boron Doped Diamond, xi
by-products, vii, ix, 2, 12, 22, 23, 24, 43

C

cell death, 1
cell size, 26

certification, x, 40
chemical, 1, 10, 16, 28, 30, 34
chemical reactions, 28
chemicals, 9
China, 42
chlorination, 23, 45, 46
chlorine, vii, ix, xi, 1, 2, 10, 11, 16, 17, 21, 22, 23, 24, 25, 26, 29, 30, 35, 36, 43, 44, 46
chromatography, 10, 12
chromatography analysis, 10
compensation, 9, 15, 31
competition, 17
complexity, 24, 27, 39
compounds, 14, 30
conductivity, 9
constituents, 13
construction, 2, 26
consumers, x
contact time, 22, 23
corrosion, 5
crystallization, 9

D

DBP, 44
decomposition, 13, 45
dehydration, 28
desorption, 28, 30

Index

detection, 10
diffusion, 16, 27, 30
Dimensionally Stable Anode, xi
disinfection, vii, ix, 1, 2, 6, 22, 23, 24, 28, 39, 41, 43, 44, 45
dissolved oxygen, 13
distribution, 16
dosage, 23
drinking water, vii, ix, 5, 6, 8, 9, 15, 16, 18, 19, 20, 22, 23, 24, 26, 29, 34, 36, 39, 41, 42, 43, 44, 45, 46

E

electrochemistry, 1
electrode material, 2, 3, 10, 17, 26, 29, 33, 39, 44
electrode surface, 1, 13, 14, 21, 29
electrodes, ix, 2, 5, 7, 8, 9, 14, 17, 21, 25, 42, 43, 44
electrolysis, ix, 1, 2, 5, 6, 10, 16, 21, 22, 23, 24, 25, 26, 27, 28, 30, 31, 32, 36, 39, 41, 42, 43, 44, 45
electrolyte, 7, 32, 44
electron, 13, 27, 28, 29
energy, 5, 45
energy input, 5
environment, 42
Environmental Protection Agency (EPA), xi, 6, 23, 43
equipment, 12
evolution, 33, 43, 44
examinations, ix, 23, 24

F

food, 43
formation, vii, ix, xi, 2, 3, 5, 6, 8, 9, 13, 14, 16, 17, 19, 21, 22, 23, 24, 26, 27, 28, 29, 30, 31, 32, 33, 34, 35, 36, 39, 40, 41, 42, 43, 44, 45, 46
formula, 36
France, 42

Free Available Chlorine, xi
Freud, Sigmund, 6

G

geometry, 39, 44
Germany, 9, 42
graph, 16
graphite, 2
growth, 45
guidelines, 5, 22

H

halogen, 24
hardness, xi, 9
health, x, 6, 43
health risks, x
housing, 8, 9
hydrogen peroxide, 13, 30, 35, 37
hydroxide, 10, 29, 30
hydroxyl, 13, 14, 29, 30, 34, 37

I

ion transport, 16, 21
ions, ix, 21, 22, 30, 35, 42, 43, 45
iron, 5
issues, 26

K

kill, 1, 2
kinetic constants, 40
kinetic studies, 44
kinetics, 34
Korea, 44

L

lead, 1, 5

Index

M

magnitude, 18, 28
majority, 2
malignant growth, 45
mass, 21, 28
materials, 1, 2, 3, 5, 7, 14, 17, 29, 33, 39, 40
matrix, 12
matrixes, ix
mercury, xi, 9, 15, 31
microorganisms, 1
migration, 16
milligrams, 22
molecules, 34

N

Na_2SO_4, 9
NaCl, 6, 9, 15, 16, 43
niobium, 9

O

OH, 13, 14, 29, 30, 34, 35, 36, 37, 40, 42
overlap, 10
oxidation, ix, 5, 13, 14, 16, 17, 25, 27, 29, 30, 31, 32, 33, 35, 37, 43, 44, 45
oxide electrodes, ix, 14, 17, 21
oxygen, 13, 14, 16, 30, 33, 43, 44
ozonation, 36, 45
ozone, 21, 24, 33, 36, 37, 44

P

parallel, 8, 27
pathogens, 2
perchlorate, vii, ix, xi, 3, 5, 6, 8, 10, 12, 13, 14, 16, 17, 19, 21, 22, 23, 24, 25, 26, 27, 28, 30, 31, 32, 33, 34, 36, 39, 40, 42, 43, 44, 45
peroxide, 13, 30, 35, 37
pH, 9, 10, 11, 30, 34, 44
plants, 43
platinum, 2, 5, 9, 14, 17, 33, 44, 45
polarization, 15, 16, 33
polymer, 44
preparation, 45
probability, 14, 21
producers, 2, 40
project, xiii
protection, 5, 42
public health, 43
purity, 9

R

radiation, xi, 42
radical reactions, 28, 36
radicals, 13, 14, 16, 29, 30, 33, 34, 35, 36, 37, 40
rate determining step, xi
reaction mechanism, 5, 28
reaction rate, 16, 27, 28, 34, 35
reaction time, 19
reaction zone, 27, 40
reactions, xi, 13, 14, 16, 21, 28, 29, 32, 33, 34, 35, 36, 37, 39, 40
reactive oxygen, 44
reactivity, 14
regulations, 2
relevance, 21, 23, 42
researchers, 27, 28
risk, 26, 40, 43
risks, x
routines, 40
ruthenium, 11

S

safety, 6
Saturated Mercury Sulphate Electrode, xi, 15
scope, 1, 15
silicon, 5, 8

sodium, 2, 10, 43, 45
sodium hydroxide, 10
solution, 1, 2, 5, 10, 11, 12, 15, 21, 22, 23, 24, 28, 29, 31, 32, 39, 45
species, 5, 13, 16, 17, 19, 21, 27, 29, 30, 32, 34, 35, 44
spectroscopy, 10
Standard Hydrogen Electrode, xi
state, xi, 5, 25, 27, 30, 35
storage, 5, 23
substrate, 29
Switzerland, 46
synthesis, 5

T

tanks, 5
technologies, vii, ix, 1
technology, 40
temperature, 8, 9, 18, 19, 20, 21, 31, 32, 39
titanium, 2, 5, 11
transformation, 25
transport, 16, 21

treatment, vii, ix, 1, 2, 5, 6, 7, 17, 40
treatment methods, 2

U

UK, 44
USA, 6, 40
UV, 9, 10

V

valve, 8

W

water, vii, ix, xi, 1, 2, 5, 6, 7, 8, 9, 13, 15, 16, 17, 18, 19, 20, 22, 23, 24, 26, 29, 30, 34, 35, 36, 39, 41, 42, 43, 44, 45, 46
water quality, 23
WHO, 5, 23, 46
workers, 1, 5, 6, 33, 36